The Everything Machine

Gabe Bentz

Contents

Introduction

Whenever most people think of robots they imagine the Terminator or some other movie representation of a robot. But few people really understand the technology and the abilities of these artificial organisms. Even people within the technology sector themselves do not fully understand the diversity and complexity and even simplicity of these machines. I decided to write this book to give an introduction to robots where a layman can take the material to a party and have a decent conversation with someone about it.

The chapters of this book will outline many of the various types and applications of robots that have been defined thus far. Then we will explore what robots can or could do if someone would only create it. And for the truly ambitious, we will talk a little about the philosophy of robots on a purely conversation and hypothetical level. Basically, we are just going to talk about robots for a while.

This book is really a collection of essays. So, though they may build on each other a little bit, each section is really meant to stand on its own.

If you are new to robotics this will give you a good introduction to the technologies that exist and even if you are a veteran you may find something you did not previously know about. Either way I hope you enjoy.

What is a Robot?

There was a reason this book was called "The Everything Machine." We will get to that in a moment. For the time being let me explain from a technical perspective what a robot is.

At the most basic level a robot is a machine that can sense the world around it, make a decision, and then take physical action, like swimming or grabbing, based on what is sensed.

In order to sense the world robots use a wide variety of sensors and technologies. They use infrared light, ultrasonic waves, laser beams, bumpers, and normal cameras to see their world. But these sensors just give a huge amount of data in the form of electrical signals. These signals are then analyzed and conclusions drawn by the brain of the robot. Today that brain is normally a binary computer which uses millions of transistors, which are just non-mechanical switches, to make a decision and command an action. The actual bodies of robots can be any machine. Robots can have arms or wheels, they can fly or swim, and they can be powered by a gas engine or a fission reactor.

That is the basic definition and outline of robots but let's go a little beyond that. Based on what I just said your 1950's thermostat is a robot. It senses the temperature of the room, decides whether to turn off the heat or turn on the heat and then does it. But that "decision" is based on how a piece of metal curls in response to heat but is that really a "decision." Your car may "decide" to lose a wheel when it senses a bearing has gone out from too much driving without a service. A clock "decides" to add a second after is senses that a second has gone by. Are you starting to see the problem with the definition?

A robot is not some piece of clockwork that simple reacts when a button is pushed. A robot is not a machine that reacts by breaking down. It is not a machine that simply repeats some basic command that it was given. A robot is really more of an artificial organism. A creature whose muscles are motors and whose synapses are transistors. That is what a true robot is. A robot is an animal created to perform some task. It has a behavior and basic rules that it follows but it is able to change and develop based on its environment. That is what a true robot should be.

But, most of the machines that are considered to be robots are not so elegantly organic. They are just highly complicated clockworks which simply use computer code instead of wheels and gears. So for most of this book a robot will be considered a machine that uses a computer to control a physical body inside of the real world. A robot is an extension of the virtual world into the real world. We'll run with that for the time being. In reality the definition of a robot is like debating the philosophical truth of something being conscious, there are many facets to it and there is no black and white definition found in logic. That should give me the cop out needed to continue with the book.

Now, a moment ago we left the idea of "The Everything Machine" to consider life. If one considers robots to be organisms then you only have to look at the diversity of life to appreciate the diversity that robots can achieve. In fact, robots can become more diverse than the life on earth because they are only limited by the imagination of the creator, instead of mechanisms like evolution and genetics. Robots can fly, run, walk, swim, or roll. They can be as large as a building or as small as a cell. They can live in the dirt or in space. They can be made of biological cells or cold steel and plastic. When a machine or organism can be so flexible how can its capabilities be limited?

Many people fear the ability of robots to do anything. Many believe that they will lose their jobs because robots can do anything. This is partially justified and is another gray issue surrounding the industry, but it is not the physical capabilities of robots that people should fear.

Machines have long replaced human jobs. Steam and organized factories replaced skilled workers. A machine can be created to punch out car parts or clock pieces. Robots are just machines. Even if we made a perfect android it would not replace humans any more than any other machine ever has.

This is where robots are different from any other machine. Every other machine was simply a replacement to labor. All the machines of the industrial revolution replace muscle. The cotton gin killed the jobs of the people who would pick seeds from cotton (generally slaves but we will ignore that in order to make the point). The sewing machine replaced the many women creating and sewing new clothes. But these machines were simply a more efficient use of physical energy.

What is threatening about robots is that robots have brains. Currently these brains are highly limited but they are brains. Robots do not simply replace the labor of people but they could replace the ideas and intelligence of people. This is what people can fear. Robots are ideally a perfect artificial facsimile of humans.

But that is all still some time away. Few have to worry about an android walking in and replacing you at your desk for now. Today robots have other strengths that could lead to the kind of increase in productivity that came during the industrial revolution. Like any machine robots are great at doing repetitive things. And since robots are a little smarter today than normal clockwork machines (though not as smart as people) they can do fairly complex repetitive things.

In the movie WALL-E there is a sequence where the audience sees this little robot collect trash, compact it into cubes, and stack those cubes. Over several shots the audience sees that the robot is creating a skyscraper of trash cubes. It takes a long time but the little robot does not want to do anything else. It was created to compress and stack these cubes of trash over and over and over and over and over again until his battery goes dead. A clockwork machine can't do this but a simple robot certainly can. A basic artificial organism, as we defined robots to be earlier, can certainly perform such a task. Can you imagine a human doing something so menial and repetitive for their entire life without so much as a coffee break. It would drive them insane or wear them out. A robot does not become tired they simply break down. And a robot has an unrelenting determination, in completing a task, which no human can match.

As I said before robots are dumb. But there are many things that dumb things can do. Look at ants. No single ant has any idea how to design an ant hill and yet together with a few very basic ideas they create complex architecture and society. Robots can do the same. Today robots can be put to work performing tasks where their determination can be put to good use. They could combat invasive species one plant at a time. They can clean oil spills one drop at a time. They can monitor borders night and day without ever falling asleep.

Today people need not fear that robots will take their jobs. Robots are still too dumb. But imagine jobs that have been abandoned as the standard of living has grown. The minimum wage it too high in America to pay to have rocks picked on the side of the mountain. So a robot can be created that can do the work, which would have otherwise gone undone, for less than the cost of a human. So you see that humans can improve their quality of life and leave certain tasks behind and robots can fill the void that is left. Robots will not replace humans they will simply be given new jobs that were either undone, forgotten, or never thought of.

That ends the outline of what robots can do and what they could do. Later in the book we will discuss in more detail the effects on society of a few robots. The next chapters will talk about the different types and applications of robots that are currently in existence.

How Robots Think

No book on robots would be complete without at least a brief discussion on the technology that makes such thinking machines possible. The computer.

Robots think through the use of computers. Computers are glorified number crunchers. Many people think that since a computer can make a character on Grand Theft Auto talk to you and move around the screen that they are highly complex machines. Not to marginalize computers, but they are incredibly simple.

At the heart of every computer are billions of transistors. Transistors are just switches that store and process information by saying yes or no to whatever comes their way. When one puts a billion such switches into the same machine seemingly complex things can happen. Transistors can be likened to sand on a beach. One grain is nothing but when you pile together a billion you have a sand castle. One transistor is nothing but a billion of them gives you a video game character.

Now, transistors in binary computers today are machines much like gears in a clock. They respond in very predictable ways. So technically a robot or computer has a brain that is just a gearbox filled with a billion gears. But computer science is not that simple. Transistors store and decode information but they are not the final decision makers. The actual thinking of robots occurs in computer code. In reality transistors are like the DNA of a computer. They control things like speed and processing power and appearance but they aren't what control thoughts. In biology thought comes from cells called neurons. In computers the neurons can be considered the small programs of computer code.

For years computer code has had the capability to emulate imagination and original thought. Computer programs have written songs and made paintings. But this is due to several neat tricks performed by computer scientists. Such programs detect patterns and then emulate them. This process is called machine learning.

Machine learning is the discipline of having a computer basically program itself. Companies such as Google are highly interested in this field as it will allow Google to understand what you mean when you type in "I want ham." Normally the search engine would match those words to top sites, but machine learning will allow it to understand that you really want the computer to order a ham from the grocery store and deliver it to you. Machine learning is an attempt to give computers, and therefore robots, the ability to operate in the real world as a human would.

The pinnacle of machine learning today is what is being referred to as Deep Learning. This technique is the use of connections between information in an emulation of how neurons function in the brain. These connections combine and form patterns and pathways of computer code and information that is incredibly accurate, predictive, and robust.

In robotics, Deep Learning will allow the machine to realize that "handles" on objects are meant to be grasped. Or that a "book" is composed of pages. In the past, these types of common sense facts would have to be written as rules within the robot. The creators had to minutely define the entire world for the machine to operate in it. Deep Learning is allowing machines to learn organically like a child. So instead of creating an encyclopedia of rules that apply to everything inside the computer brain of the robot, today people are working to have the machine make assumptions based on its experience and limitations. This type of technology will allow robots to operate within homes and businesses, where the environment is constantly changing, and learn new skills that were never a part of their original design.

Computers are still just a billion switches. However, using that DNA, engineers and scientists are creating brains and artificial neurons that will allow machines to think like humans, at least from a mechanical standpoint.

A Brief History of Robotics

Robots are actually a very old idea. The name comes from the Czech word "robota" which refers to drudgery or servitude. The word "robot" was first used in the play "Rossum's Universal Robots" (R.U.R.) in 1920. In the play the robots eventually rebelled against their human masters and killed them.

Another book known to influence the field of robotics and artificial intelligence is Mary Shelley's "Frankenstein." In this book Frankenstein's creation turns against him and leads to great devastation within the Frankenstein family. It is an insight into the creation of intelligences outside of natural means.

Leaving the literary world, automata of some kind have existed for hundreds of years. Leonardo Da Vinci even worked on robots. He created automatons for entertainment. Such machines could not perceive anything (though it is argued that robots still do not). And their actions, though as complex as writing a sentence, were hardwired and could not be changed. These machines could write a sentence but they couldn't tell whether there was any paper. They could be likened to music boxes with arms.

Robots that could almost be considered artificial creatures began appearing in the 1950's and 60's when semiconductor technology made them possible. Many of these robots were little machines called turtle-bots which were created by psychologists in order to study intelligence.

The first truly practical robot system was Unimate, an industrial robot arm. This robotic arm was designed in the

1950's by George Devol. It was implemented in GM auto assembly plants to act as a welder in 1961.

While Unimate is what many consider to be a modern robot, it had electronic memory, it was articulated, it had sensor feedback, it was still basically a clockwork mechanism. What made it different from any previous clockwork machines was that it could be programmed. The clockwork could be changed by simply loading some new 0's and 1's. This is what made the Unimate valuable because instead of having to retool an entire factory for a new product the robots just had to be reprogrammed. That capability moved robots ahead of other basic machines. Robots can perform many tasks by just reprogramming them. Clockwork machines may only perform one single task.

But, Unimate really isn't a full robot. It couldn't make decisions it simply executed actions. The fuel injection system in your car is equally intelligent.

Since then industrial robots, based on Unimate, dominated the market. In the 1980's when computing was just beginning to become widespread several start-ups attempted to create home robots. All of these robots failed. They simply couldn't do anything useful and the user had to program the robots using computer code. They were not good products, especially when people expected Rosie the Robot.

In the 90's personal computers and the internet stole much of the talent that might have studied robotics. Algorithms which worked in the cyber world behaved better than algorithms in a robot in the real world.

But in the 2000's robots got a boost. A company named IRobot, founded by Rodney Brooks and several MIT graduates, created a robotic vacuum cleaner. The Roomba could clean a house decently and became a successful product. IRobot soon became one of the largest robotics companies that existed outside of industrial systems.

Also in the 2000's, computing began to really become powerful and products like the smartphone and the cloud have allowed robots to become exponentially smarter than they ever could before. Unlike previous decades a person can now go online and purchase a robot that can perform tasks in the house or in the office. And the technologies for these robots are actually quite impressive. Though what happens in the next few decades will make the robots of today seem absolutely Stone Age.

Telepresence

Telepresence is technology that has been gaining traction primarily over the last 5 years since internet speeds have increased substantially.

A telepresence robot is an RC car with a camera on it. That is fundamentally as complicated as they get. The main purpose of telepresence has been to reduce travel costs for businesses by allowing someone in New York to drive this little robot around in San Francisco, thus saving a plane ticket. Most of these robots look like a TV screen on a stick. The screen shows the face of the person driving the robot in order to increase the "presence of the person." A telepresence robot establishes a legitimate physical presence with people across the globe without having to actually travel there.

Now, most have argued that telepresence is a little over glorified. This stems from the fact that everyone has video chat of some kind on their phone. Why not simply have someone at the other end move the phone around the factory and give you a tour through Skype.

That is a legitimate point but the first issue is that it requires two people to work. It also eliminates the independence of the person in the phone. Telepresence robots give a rough illusion of a second body that the person has full control of. And many believe that the physical presence and general interaction is better than traditional video conferencing.

Some technology companies agree. Google is backing a telepresence robot company called Double. The robot that Double makes balances on two wheels like a Segway and instead of having a bunch of computing onboard it simply interfaces with an Ipad for the screen and camera needed. The Double is an Ipad on a stick.

IRobot has designed a much beefier system called Ava 500. This robot is still just a screen on top of a mobile base but it has a much thicker structure.

Business has been the primary application for telepresence robots, due to their ability to save travel costs. But they are also just beginning to come into the home. Small robots which you can plug a smartphone into and they become a robot are starting to gain ground. One such product is Romo.

Romo is a tracked robot that uses your smartphone as its brain. When needed that phone can also serve to turn Romo in a robot that allows parents to check on the kids or the pet.

Home robots will likely become more popular in the coming years as the population of the US and other countries age. Telepresence will allow family members and nurses to visit the elderly more often, perhaps allowing the old folks to remain independent longer.

Telepresence robots may also serve as an intermediate step to truly artificially intelligent machines. Instead of having a program control a machine perhaps workers in India may drive these robots stocking shelves or weeding corn rows in America.

Ultimately telepresence may become advanced enough that instead of travel people inhabit an android body and walk around just as if they were human. This concept was portrayed in the movie "Surrogates" starring Bruce Willis.

But there are some technological hurdles to make before many of these teleported workers can become reality. The first is in the intelligence capabilities of these robots. As I said telepresence robots are basically an RC car with a camera. What makes them robots is the fact that the system has just enough smarts and sensors onboard to help the human drive it. It is actually difficult to drive one of these robots without practice. You end up running into chairs or cats or toys without even knowing it. The robot side of the system is there to act a copilot. Since the person driving has very little situational awareness of what they are doing, the robot compensates and might pause to let the cat go on past even though the human is pushing on the gas pedal.

For the "Surrogate" concept to work the interface has to be improved. Right now people basically use a PC with arrow controls to drive the robots. A robotic replacement of the body systems will have to learn to read brain signals and intent and then the robotic side will make sure that the body remains balanced and still dodges the cat.

When people think about collaboration between robots and humans, telepresence is really a prime example. In no other situation does a human and a machine have to work together so perfectly for it to operate correctly. The only thing which requires greater integration is cybernetics.

Personal

What would a robot in your house look like? How would you imagine it to look? Would it be square like the lovable Rosie from the "Jetsons" TV animated sitcom? Or would it be more like some kind of C3PO that carries a tray and waits on you hand and foot. Or perhaps you imagine the home robot to be a swarm of little fellows that rush out of their nooks and crannies when needed. One mops up a spilt drink, another retrieves a new one. I am sorry to say that none of these exist today.

The closest thing that anyone today can get to a personal robot is the Roomba vacuum cleaner. Though there are other companies in the robot vacuum market, IRobot's Roomba is the leader and most of the others are functionally similar.

The Roomba is a little robot that scurries around your house twice a week and vacuums your floor. This little robot looks like a hockey puck that met an enlargement ray. It has a small bumper that lightly contacts the furniture and then leads the robot around it. When the job is believed, by the little fellow, to be finished it returns to its charging bay where it sits and charges its battery until it is scheduled to vacuum again.

What I just described is very similar to one of the robots in the swarm that we imagined a moment ago. In fact this is what IRobot is going for. They want to create multiple single-purpose machines which work together to perform the functions of a single robot butler.

The Roomba is a very simple machine, from a robotics standpoint. At its most basic level when it hits something it turns, and it's dragging a vacuum. That is all it does. But it does it well. Mundane and repetitive tasks like vacuuming are something that robots are very good at. And when a robot only has to do one thing, for example bounce off furniture, it can be made very cheaply. Thus the Roomba has existed for over 15 years and is still the leader.

But what about Rosie or C3PO. If you ever go onto YouTube and really look for it you can find several demonstrations of robots cooking or even working together to make a few pancakes. But what is odd is that the robots making the pancakes take 15 minutes to just find the batter. Then they take another 15-20 minutes to stir it up. It almost appears that they are moving in slow motion. This sloth is due to the robots having to think about every action. It is incredibly difficult for robots to work in environments that are unknown or disorganized. The creation of order out of disorder, say getting the milk from behind the lemonade in the fridge, is phenomenally hard for robots. While a human automatically knowns that you simply move the milk aside, a robot can't tell how much the milk carton weighs or even where it is without a huge amount of number crunching and experiments.

One would expect that this problem will be solved in a few years as Moore's Law does it's work and doubles the speed of the robots' noggins'. But all of that work was simply to get a box of pancake batter out of a cupboard or the lemonade out of the fridge. We haven't even begun to work on taking it to the dining room and placing it on the table.

But all is not lost. Robots are beginning to learn. And unlike humans when one robot knows something all the other robots can know it too. Robots can have a hive mind where when one learns to grab the milk he can teach the others to do it also, instantaneously.

Much of robotics, and even personal robotics, in the coming years will not be in figuring out an algorithm that will let a robot do everything. But it will be to learn how to teach a robot to do everything. Roboticist need to get robots out of the lab and get them to begin learning.

This is being done. That is why Deep Learning was discussed. There are many programs and robots which scour the internet trying to learn that tires go on cars and that apples are different from oranges. But this is taking a long time and they still do not compare with humans.

So what will an actual personal robot look like in the future? At the moment it appears that we may be headed toward the swarm of single tasks robots. Robots who are basically appliances. One vacuums, one dusts, one takes clothes to the washing machine, etc. The robot butler is lagging behind.

But say that a robot butler was created today. In order to be affordable it would have to have few motors. So it may have one arm. And so it can operate in your home, a technician would likely have to deliver it to your house and calibrate it to items like the colors of your home. He could teach the robot that the dining room is red and the kitchen is yellow, a terrible color scheme, but at least the robot knows where he is with very little effort. Then the technician would lead the robot around to different areas so that it becomes familiar with your home. Ideally you would also have the house arranged as you would like it to be. Then the robot has a foundation to work from. If a shirt is ever on the floor he will know that it does not belong there and takes it back to the laundry.

When the technician leaves it will be up to the owner to train the robot. The owner will treat the robot almost like a dog. If the robot does something he shouldn't a firm "No" will teach the robot the lesson. Slowly the robot will settle into its new home and will work to keep it tidy. It should even be able to grab something like a Swiffer and clean your floors. That is what the robot butler of the next ten years will likely look like.

Industrial Robots

In every manufacturing documentary ever, one will see an assembly line with dozens if not hundreds of robots working on cars. Sparks will fly as the narrator explains how these machines have revolutionized manufacturing and made the threat of job replacement by a robot real.

However, most of these industrial robots are not really robots. They are simply highly complex clockworks. They execute the exact same motions in the exact same order no matter what, just like their father Unimate. If a person steps into the fray he will be mulched and welded as if he was one of the cars. The robots are just large machines that can only be retrained by a specialized engineer.

Now these robot arms are highly interesting. The control and the symphony of movements is an engineering marvel, but so is a skyscraper. It doesn't mean that it is a robot.

So there are giant arms that building cars. Fine. But there are new industrial robots on the block that are just beginning to gain traction.

Rethink Robotics is a company located in Boston, Mass. Their claim to fame has been a collaborative robot they call "Baxter"

Baxter is a two armed industrial robot that can be trained and retrained to do something by the average person and is safe to work around, unlike his behemoth cousins.

Baxter is probably most highly defined from his cousins in appearance by his face. Most industrial robots are a huge hydraulic arm and nothing else. Baxter has a screen for a face, with two eyes that show what the robot is paying attention to at any given time.

Baxter is called a collaborative robot. The term refers to a robot that humans can interact with safely and easily. Ideally every robot would be collaborative. But in Baxter's case the definitions goes a bit further. Baxter is meant to not be a replacement to people but an assistant. A robot that is wheeled in to some point on the assembly line and grabs parts from a conveyor and then hands them to a human worker. Collaborative means that the robot and the human have an almost symbiotic relationship. The robot is too dumb and slow to assemble parts and the human is too fast and smart to just move a piece from one place to another. Each works in their own ideal world. The robot doing highly repetitive and boring work, the human doing highly articulate work.

Since Baxter is also easily trained and retrained he is an ideal tool for small batch manufacturing where what is made and how it is made is often changed. Traditional industrial robots would need on the order of weeks to be reconfigured. Baxter is simply wheeled into place and a worker takes Baxter through the motions and then Baxter learns what he is supposed to do.

Several other companies are working on robots similar to Baxter. Universal Robots makes several arms that have the same capabilities as Baxter. But of all the robots that exist Baxter seems the most friendly. He also meets the Hollywood expectation of what a robot should be. He looks like a "he" not an it. When that happens a robot is very close to being truly a robot in the eyes of the user. Otherwise it is just a machine.

In the future industrial robots will all likely be made to be easier to train and configure. This necessity is coming from the increase of niche manufacturing made possible through technologies like 3-D printing. Also, with the very basic kind of work that has to be done by industrial robots the training of them should not be so difficult.

On another point. Manufacturing is changing. Engineers are becoming less and less needed in the design phase as computer design programs learn to generate efficient and effective designs through algorithms instead of genius. When computers have reached the point where they design the product and then robots manufacture the product humans will simply be the bottleneck in the process. So while for the short term collaborative industrial robots may be effective there will come a point when having humans train and work with robots will be far more inefficient than simply having computers tell the robots what to do.

Military

Like many technologies robots have had a very good early following in the military. Drones and unmanned vehicles have had significant boosts from organizations such as DARPA and the Pentagon.

Military robots is a very broad term. The military is almost like a society so robots fill many different positions within it, from carrying the bullets to actually firing them. A military robot is basically any number of the other robots discussed in this book which has been ruggedized to meet military specifications.

With that being said, many of the robots that have been actively used in the military are not true robots at all. None of them do much thinking. The IRobot PackBot is a bomb disposal robot which is little more than an RC car with an arm and tracks. The machine may look like a robot but it is just a complex and rugged toy.

The Predator drones and other such planes are a step above the PackBot. These planes can operate alone. They can survey an area and even identify targets but if something highly complicated arises then a human pilot in some bunker or warehouse has to step in and take command. Such manual control becomes necessary when a trigger is pulled or some complex aerial maneuver must be made, often takeoffs or landings, though robots can now do these alone also.

The military was highly interested in autonomous vehicles for supply chains. Such vehicles would be more efficient than people and would eliminate the danger of minimally protected convoys being points of abduction.

Robots are also starting to evolve to help carry the load of soldiers. Legged robots like Alphabet's Boston Dynamics Big Dog is a four legged walking robot that is meant to be a packhorse for the Marine Corps. Unfortunately, this project has hit resistance because of cost cutting and the Marine Corp not seeing vital benefits from the systems during testing. The 42 million dollar development of Big Dog's big brother, Alpha Dog, was canceled in 2015 because the gas engine that powered the robot was too loud.

Now, on to the very serious stuff. Do military robots shoot people? Yes. Predator drones have taken out targets and there have been ground robots outfitted with machine guns which have patrolled areas, there are even turrets on navy ships which can identify an incoming target missile and shoot it down. For most of these systems humans have the final say. From what is publically available no robot is given carte blanche on pulling the trigger. This "human in the loop" requirement has been established because robots are not good at dealing with complex environments. Suppose a robot was told to shoot anyone carrying a gun who was not wearing an ally uniform. That can be done. But the robot will not differentiate man from child. Some would argue that robots might be better than people because they lack emotions and would not allow a child to kill someone. But robots are still not reliable enough to differentiate friend from foe or to make judgment calls in very complex situations.

Robots will become smarter and will eventually be able to identify and discern targets as well as people. When that day comes the fundamentals of war may change. Suppose that instead of sending armies of men to fight each other there were only armies of robots. How would such wars be won or lost? War would cease to have any emotion and would rather become a matter of commerce. Whoever loses their army first loses the war.

In the case of robots fighting against humans there will be no contest. Humans would not be able to sustain against the unrelinquishing battering that an army that does not eat, sleep, or even pause to hear an instruction could sustain.

Robot armies may eliminate some isolationism in the world. When no human lives can be lost incursions by larger powers may become more frequent as there is less internal backlash from families that have lost members.

Robot wars can only occur in two extremes. They will either end in victory for the richest side which can sustain manufacturing of their army or they will end in victory for the side that decides to attack humans as well as robots.

An artificially intelligent army would be almost equivalent in weapons technology to that of the atomic bomb. Robots may be slower but they can be as thorough and devastating as the seconds of a nuclear blast.

Drones

Today you can hardly get through the news each day without coming across something having to do with drones. Drone is a poor word that has been adopted by media and marketing. Technically most of the toys and vehicles referred to as drones are unmanned aerial vehicles (UAV). A "drone" is technically any unmanned semi-intelligent machine. So an RC car could be a drone. But enough with nerdy semantics.

Drones have gained popularity for basically two reasons. They fly, which is cool, and they can take pictures, which shows how cool they are to your friends.

Most drones being created today are built around a helicopter like design which uses multiple rotors to lift and tilt the craft. Such designs have only become feasible when computers and control technology evolved far enough that they could be housed in such a small package that it could fly.

In a drone system a human is generally flying the craft but the software and sensors on the vehicle itself make sure that it remains stable. Basically all the human does is tell it to go forward, backward, right, or left, then the machine figures out how to do it. That being said, not all drones can be taken off the shelf and flown with no experience. There is a learning curve because the drones are not smart enough to argue when you command it to do a backflip incorrectly.

From a practical standpoint drones are a useful surveillance and surveying tool. With enough software and battery life these become robots that can monitor borders, track crop growth, and watch traffic for much less than a helicopter reporter. But aside from being an eye in the sky drones have not found much other use.

Amazon and Google are two of the largest companies trying to make drones useful as something other than a camera platform. Amazon has begun testing of a drone-plane hybrid which can take-off and land like a helicopter but then fly like a plane. The application of such a system is in package delivery. Amazon wants to deliver a pair of shoes the day you order them. This has been achieved on the engineering side but there are regulatory issues slowing the implementation in the US.

Since drones are known to function as mobile cameras privacy issues and air-rights have arisen. There is also the problem of one running low on battery and falling out of the sky on grandma's house, or worse, her head. Liability and right of way are still being sorted out at the writing of this book.

The delivery and transportation capabilities of drones will likely see them used as couriers in metropolitan areas. And they will likely continue to be an inexpensive aerial camera system. On the humanitarian side there have also been efforts to use drones to deliver medicine and medical samples to and from areas with little transportation infrastructure.

Space

Sending a rocket to space is expensive. What is even more expensive is putting a human in that rocket. When every pound matters, tons of life support and room to move really eat into a budget. Robots, on the other hand, can fit inside a vacuum- sealed box and never even feel stiff on the way out.

NASA has been using robots to explore space primarily because of the cost basis just introduced. Robots are also expendable and can sometimes operate longer than humans. The Voyager probe is a robot which is likely going to continue its mission for another few decades beyond the solar system.

When discussing space robots is it is almost entirely a comparison to humans. Humans are more discerning, robots are more efficient. Humans get more public interest, robots return more data. The argument has raged for years about whether exploration should be done by man or machine. The result has come to be a symbiotic system where robots blaze the trail and make sure that a pitfall is not waiting around the corner for the man. Robots are the scouts of space studying as much as possible with their limited capabilities.

The pinnacle of space robots today is likely the Curiosity rover on Mars. This six wheeled SUV-sized artificial explorer is powered by a small nuclear reactor and was meant to search for life on Mars.

Curiosity was lowered to Mars at landing by a crane that used its rockets to hover above the Martian surface as it lowered the rover to the ground. Once landed, Curiosity unfolded its sensor studded head and looked around the

Martian landscape. Curiosity has several labs built inside of its body which allow it to analyze Martian dirt for biological materials.

Curiosity is very much like most drones today. It has very limited intelligence. In fact it has almost no intelligence. Every one of its actions is controlled from Earth by a group of NASA engineers. The engineers steer it and tell it where to drill. Really all the robot does is make sure that it does not press too hard on the drill and break a bit. The robot is basically just a safeguard against human error.

Robots have dominated space exploration for several decades since the Moon missions. They have been the only explorers beyond Earth orbit. But today companies such as SpaceX and even plans by NASA are working to put people on Mars. In this situation robots will shift from being exploders to construction workers. Autonomous machines will have to be created which can clear and build space for the astronauts and colonists to land and live in. And as those people arrive the robots will work side by side with the people in order to grow the colony and even someday terraform Mars.

But robots will likely continue to be the trailblazers in space. Their lower cost and greater expendability makes them the ideal candidates. Also, since robots can be made to live in any environment they have the ability to explore so much more. Robots will be the first to see the oceans under Europa's crust and the first to travel to another star. Perhaps even the first to tread on another earthlike planet.

Smart Homes

You walk into you home and immediately the lights come on. The room is also the perfect temperature. While you were driving home your phone sent the house the message to get ready for you. Your favorite music is playing and dinner is going to be delivered any minute (possibly by drone).

That scenario is what smart houses will make a reality. A smart house is basically an intelligent machine you live in. The house learns your preferences and habits and eventually modifies itself to match you. Lights, heat, cooking, cleaning, all are controlled by a brain in the basement.

Smart houses are actually relatively simple. Today anyone can turn their house into a smart house. Products like the Next intelligent thermostat learn your habits and adjust to them. Amazon Echo gives you verbal control of smart house components. Philips has made light bulbs which can connect to Amazon Echo so that you can verbally command them on or off and set schedules. Small immobile personal robots like Jibo can connect with all of your appliances and order you dinner if you request it. And more and more products are being made to be controlled by your smartphone or each other.

Today a smart home is not really a single robot with circuit boards running through the walls but a collection of individual products and robot-like machines working together in harmony to create the illusion of a single system. In the industry this is what's known as the Internet of Things. Physical products working together and communicating to accomplish a single task.

But while today you can get gadgets and smartphone apps which turn your home into a robot these gadgets are becoming more robotic all the time. Several small start-ups are working on machines which can cook and mix drinks. Some of these are small machines like a coffee-maker others are basically two arms which reign over the kitchen. Houses may eventually be built to accommodate more technology. Smart locks and RFID tech could be built into the doors and walls eliminating physical mechanisms like keys.

As with all such technology privacy is a concern. If a house is going to truly interact with and help you it has to see you. If it can see you then does the NSA or Google see you too? The limits of the technology will have to be set by the homeowner. But it is difficult to keep the house out of the bedroom.

The idea of the omnipotent intelligence whose voice calls to you in the shower is still a ways off but all the technologies exist. You can speak and give commands. A computer can track your locations and speak back. And yes robots can cook and order dinner. But most of these technologies have not been interfaced so cleanly to create a true consumer smart home.

But there are many such homes in existence. Bill Gate's mansion is intelligent in that it identifies people within it and changes the conditions or pictures in a room to match that person or mood. Companies like Philips have built conceptual houses where every aspect is controlled from a tablet.

The benefits of such houses are huge. Energy usage will decrease as the house can turn off the AC or heater when no one is there to appreciate it, but also turn it back on so the room is comfortable before you return. The house can protect itself when you are away. It can call the cops if there is a robber in the driveway or a fire in the kitchen. If its owner is a grandmother, when she falls the house may be able to detect it and call for help. The faucets will not be left running and if they drip then the house may be able to remind you to call a repairman or just do it itself.

Smart houses are actually the full implementation of technology into our lives. When a person adopts a smart house they are immersing themselves in technology. It cannot be removed, or the battery go dead, or even the alarm dimmed at times. A smart house is literally an omnipotent presence of a robot that you trust without reserve to lock the door behind you and call the ambulance if you are hurt.

Self-Driving Cars

Self-driving cars are right up there with the home robot butler on everyone's sci-fi wish list. The difference is that self-driving or "robot" cars are significantly easier to achieve and much closer to being implemented.

Google made headlines in the first few years of the twenty-first century when it was revealed that it had several cars driving around which did not have drivers in them. And they had only had 11 accidents all of which were attributed to human-error. Google has since released its plans to manufacture a driverless car and other companies such as Apple have followed suit. But at the head of the pack is an electric car company called Tesla Motors.

Tesla is a company run by a man who has invested in spaceships, artificial intelligence, solar power, and, with Tesla, electric self-driving cars. Currently, every Tesla automobile has an auto-pilot function which allows the car to drive on highways without need for a human. However, this is not true autonomy. A human is still expected to supervise the car to ensure that no mistakes happen, but this recommendation is often ignored by the drivers. Tesla's CEO, Elon Musk, intends to have his cars fully autonomous by around 2020.

Self-driving cars have been an ideal of many technologists since computers were created. But only until this century have computers had the processing power to deal with the speed and conditions of driving.

The efforts were stimulated when DARPA created the Grand Challenge. This competition required that teams create a vehicle which could navigate a 200 mile course without any human assistance. The first year of the competition the best robot went eight miles before being decommissioned. The second year the competition was won by the Stanford team with a car which used a few lasers and a camera to see the road. That team was absorbed by Google to create its self-driving cars.

Compared to many tasks driving is not terribly complicated. Highways and roads are fairly consistent and reliable. They all have lines on the edges and centers. They have been mapped fairly well. And they are flat with a highly defined black space where the car is supposed to be. A stop sign is always red and always means stop. This is why Tesla's autopilot was the first thing to be released. Driving on a highway is easy. The trouble comes when a car is driving through a suburb and a group of kids chase a ball out into the road or a Walmart bag blows across the street. It is not easy for a computer to tell the difference between a two year old in a white shirt and a white plastic bag. Right now the car will stomp on the brakes either way. Well that is good for the kids, but what happens when the car is going 80 down a highway and bag blows across the road? That is one of the problems being solved today by Google and Tesla as they work on their cars.

But with the advent of such new technology regulation is going to have to catch up. The biggest question with driverless cars is who is responsible when an accident occurs. Say an autonomous car collides with someone driving their old pinto. Is the driverless car considered to me more infallible than the human? If the driverless car is to blame does the liability rest on the manufacturer or the owner? We certainly can't punish the car.

The ethical questions will also become far more demanding. Say a driverless car is going down the street.

For some reason there is a man crossing the street and the car cannot stop. The only way to miss the man would be to swerve and instead collide with the people on the sidewalk. The question is whether the car should be programed to value the many over the one or do something else. Maybe the car might swerve if it is Bill Gates crossing the street. Or maybe is will run the man over if he is 90 and looks depressed. These situations are going to have to be programmed into a car. Some engineer or product designer will have to decide what is right. It is a difficult set of philosophical questions which have never had to be addressed in practice because before now everything has been done by humans. A car does not have feelings and it certainly will not question its actions, and again, it won't learn any lessons when you send it to jail.

But on the bright side driverless cars will be a boon to society. They will be more efficient because they can network with each other so that traffic flows continuously. There will no longer be the hours wasted in traffic or even driving to work. Driverless cars could turn an hour of driving into an hour at the office or an hour with the family, completely.

Driverless cars will also be much safer than humans. A computer does not get drunk, it does not look at its cellphone for a text from the boyfriend. And if it breaks down it will be able to warn all the cars around it that it is going to crash so they can all start avoiding a crash seconds before it happens.

Some have postulated that driverless cars will even start to eliminate ownership of cars. Since a driverless system does not require a driver getting paid by the hour taxi services will become cheaper. Driverless systems also can be present only when you need them. So instead of driving to work and then parking your car, you could drive to work and then release your car. While you are at your desk your car could connect with Uber and shuttle people around until you want it to be back to drive you home. Precisely that scenario is why the hailing app Uber is one of the companies investing in driverless technology.

Construction

Construction is a task that is just beginning to utilize robotic systems. At the moment the "robots" used in construction are just miniaturized machinery without a human inside. But with new software and building techniques that will rapidly change.

In 2010 Husqvarna released a remote control robot that is more of a miniature excavator that the human does not ride. That particular machine is used for small-scale demolition. The hope was that the system could eventually be automated so robots could go inside of a building and gut it and then intelligent construction crews could follow it. This application is perfectly legitimate. The robots would be doing a job that is hazardous to people and most people do not want to do. But little has come of it since the human-operated version arose.

A separate company in Switzerland has created a system that can lay down brick and mortar. The robot basically is an industrial robot arm with a few tweaks that turn it into a brick-laying 3-D printer. This robot has been used in the real world to created artful brick walls since the robot can created undulations in the brick pattern that are both structural and aesthetic.

Considering 3-D printing, multiple systems are being prototyped and tested that can autonomously build full houses. These robots look like large gantries that the structure takes form within. Some of them literally eject concrete-like building material and slowly raise the walls of a structure. The hardest part of the process is setting up the robot. But they hold potential for rapidly built, custom housing at a much lower cost.

Construction robots are unique in that there is really no clear direction for them to take. They have neither endless possibility nor a straight and narrow path. It will really depend on how the industry decides to approach such automation. It is very possible that construction robots will work as small laborers on the side of a site, building frames and individual parts as needed, instead of an entire structure. But as robots get smarter they may reveal means of construction that people simply have not thought of. That fact that a computer can today take a structural design and improve it reveals that humans may have a lot to learn from robots about building things. And when robots are designing the houses they are likely the only ones that can build it. Humans may come to be too imprecise to put a brick right where it has to be for the building to stand. This is already happening. A pair of Dutch robots is being created to 3-D print a computer generated footbridge to demonstrate such creative potential. This project is being undertaken by a company called MX3D.

Farming

Besides the military no other field needs more rugged and diversified systems than farming. Farming robots range from drones that observe the health of a crop to swarms of ground robots spraying weeds.

Farming robots are the least publicized of all the other types discussed in this book. Mainly because they are boring. Which is more exciting, a tractor that drives itself or a robot dog that runs with Marines? But farming robotics is potentially one of the most vital areas of improvement as the world population grows. Robots with a green thumb are becoming more needed than ever.

Today there are few commercially available farm robots. The widest implementation of robotics is in large tractors such as John Deere. These tractors have an autopilot function. A driver drives for the outline of a field or drills the first couple of furrows but then he lets the tractor do the rest while he makes sure it doesn't drive into a ditch or over a boulder. These tractors work mainly though GPS. Once the field is outlined by the driver a boundary of latitudes and longitudes is created. The tractor can then tell where it is with respect to that boundary or the last row it made. GPS accurate enough to implement such features has only recently become commercially available.

While commercial robots have not yet gained ground in the market there are many universities working on systems. The focus has been on drones. Drones' inspection abilities make them ideal for observing crops. Flying robots are being developed which can fly over a field and identify areas with excesses or deficiencies in water. They can also observe the overall health of a crop as it comes up better than a farmer can, without a lot of effort. After all, it is easier to have a drone fly over a corn patch to see how it is doing than to walk over several acres checking.

But ground work is necessary. Weeds need pulled or sprayed and the crops need thinned. Modern farming equipment has become fairly good at performing these tasks on a large scale and whatever is left is done by manual laborers. But the large scale systems are inefficient. Water and weed spray are farming items which increasingly need to be conserved. Both are often wasted on areas that do not need them because there is no other way.

So small ground robots are being created that can drive under the crop on the ground and identify weeds. These robots can then spray single plants. And since a robot really never needs a break it can spray a million weeds in a couple of weeks. And a few of these little weed-whackers could do twice that amount, and so on.

The farming industry is incredibly efficient today. One man can do what required dozens fifty years ago. But robots can up that productivity even further.

In the future robots may also be unique as the only ones who know how to farm. Imagine urban gardens on an industrial scale. Robots can be programmed with the skill-set to raise plants and make them produce when city-slickers have forgotten those talents. With that being said for something like that to be accomplished people with rural farming backgrounds will have to transfer over to technology in order to pass their knowledge on to the artificial offspring.

Robots for Hazardous Environments

The number one use for robotics systems since their inception has been to do things that humans either do not want to do or simply can't do. It is very difficult for a human to walk into a reactor that has melted down and turn the water back on. Robots are much more resilient.

Since robots are a machine they have the supreme adaptability and durability that that offers. Robots allow us to create an ant that can explore a volcano or an alien to go into a reactor.

Today there are many robots for hazardous environments. One can replace a firefighter at a hose. Another can crawl through sewers like a snake, and, especially since Fukashima, many can work inside of a broken reactor.

A lot of the robots that perform these tasks are basically telepresence systems. They are controlled by a human somewhere. The trouble has always been that in hazardous environments control by a remote operator is too slow. These machines can be almost painstakingly slow in executing their mission.

Robots working in hazardous environments need greater intelligence of their own so that they can react to what is happening around them and not have to pause to ask permission from a person to raise a finger.

DARPA again has its hand in this area. In 2015 DARPA held a robotics challenge. Teams had to create robots which could perform various tasks that might arise on a disaster site such as drive a car, close an industrial water valve, and navigate difficult terrain. Many of the teams utilized anthropomorphic robots to accomplish the tasks.

Though some robots were quite unique. JPL created a machine which looked a bit like a four legged spider.

The future of such robots doesn't really look too much different. Besides getting smarter there is really no huge technological advances that has to occur for these robots to be better. Perhaps deterioration of electronics and other parts may be improved but these have more to do with material science or force fields than robotics.

Androids

Technically androids can refer to any robots with a human-like form. That is, any robot with a face, two arms, and two legs. But since most conceive of an android as a robot that looks exactly like a human, let's go with that.

Androids are one of the most mechanically complex forms to make. The human body is a beautifully made machine. The feet of a person are a marvel. The hands of a human have not been replicated by engineers yet. The body of a homo-sapien is so elegant, condensed, strong, and complex that scientists and engineers are working now to just understand it. The simple act of walking, as done by a human, is still unquantified so that robots can do it. While it may take you only a few calories to walk across the room it takes most robots the equivalent of several hundred to thousands. That is why robots run out of battery in 5-8 hours and humans can run marathons.

Androids that have been created that match the human body have generally been created as a means to study the human body. A robot with a human skeleton but electrical muscles can allow physiologists to study how the body parts of a human interact. Creating such robots also allows for the advancement of the mechanics of robots, but this has begun to be taken over by pure simulation.

Androids with human-like faces have existed for some time. Japan has several androids in existence that are indistinguishable from their creators. The trouble is that illusion only lasts as long as the android is just sitting there. When it smiles, talks, or turns its head the fact that it is a robot is immediately clear. And these androids also can't move. They are fixed in place, on a chair or a couch or some other relaxing piece of furniture.

But what is fascinating is that these androids have sparked a discussion about a part of human psychology. Androids look human but features about them seem out of place and alien. For example where you see the face of a Japanese scientist but its movement is all wrong because it is an android. Your brain attempts to classify the thing in front of you and has trouble because it should be human but you know it is not. This paradox is called the "Uncanny Valley." It is reference to when our view of the world is shaken or something doesn't fit in our minds' models, so it looks "creepy." Horror movies use this to merge fantasy and reality and scare you even more. Robotics has been challenged by this because scientists are still trying to figure out when a robot falls into the uncanny valley and stops being cool or interesting and simply becomes weird.

But as cosmetic design and facial expressions of robots improve androids will be able to walk among us. But that walking part will likely be harder than appearing human.

Nanorobotics

Nanobots. The ultimate weapon that appears in movies. These microscopic machines have been depicted as being able to decompose tanks within seconds or congeal into an android. Unfortunately the reality is a bit less glamourous.

Nanobots are far from being the intelligent swarms that people think of. The most practical micro-robot implementation has been done at ETH Zurich, to perform eye surgery. The robot is more of a tiny kite that is inserted into an eye. That kite is then blown around and controlled by an array of electromagnets around the patient's head. The system is still in development and will likely be for some time, but it is the only real micro-bot that it is small enough to go inside of the body and perform practical work.

If you imagine nanobots as only being molecule-size the closest thing is an inchworm that was created at Dartmouth College. This robot is 250 micrometers long and 60 micrometers wide, so they are about as wide as a human hair. This little critter can crawl across a flat surface by dragging itself as an inchworm would do. While this robot is much smaller than the ETH system, almost everything is on the robot. There are no external magnets pushing it around. But it is controlled and powered by an array of electrodes within the surface that it moves on.

A robot made of just a few atoms does not exist. Currently, scientists can only make a few basic components, like gears out of atoms. While impressive it is still a long way from robots.

Nano robots offer the potential to cure diseases and build or destroy amazing things. Machines the size of atoms do have the potential to break down large material things like they do in the movies.

But here are the problems. Working in the nanoscale all of physics change. Moving though water when you are the size of a cell is much harder than when you are macroscopic because surface tension becomes a bigger deal. Or when you are as small as the inchworm robot surfaces become very sticky so it is difficult to move. And when something becomes as small as atoms and molecules all normal physics breaks down, that is an area that it not very well understood.

And if you can understand how those microscopic or atomic worlds work the challenges arise of actually building the robots. New methods and tools of manufacturing have to be created that operate at that level. For this reason some nanotechnology researchers are exploring nature for example. They are finding it might be easier to program a virus to produce nanobots that to try and construct the nanobots themselves.

Making a microscopic robot submarine is a very difficult challenge that really requires a better understanding of the fundamental sciences themselves. It is not just an engineering or implementation problem like many other robotics fields.

Prosthetics

Thus far we have spoken about the many applications and current technologies of robots as they exist as a species alone. But now we will take a small step back and explore how robots and humans are truly beginning to bond symbiotically to create cyborgs.

A prosthesis is any mechanical apparatus which can be added to or replace some part of an organic body. By that definition your smartphone is prosthesis. And though our mobile devices have become extensions of ourselves I am going to focus on the latter part of the definition. Basically the replacement of body parts.

The earliest prostheses were very simple wooden carvings of toes by the Egyptians. The peg leg and the hook are the most common prostheses that people recognize from the last few hundred years. But prosthetic technology began to take a jump after the Civil War when many men returned home without limbs. This was when a prosthetic arm was created that could grasp objects by the user pulling a cable through the use of a shrugging motion. The system remains in use even today.

But, with the conflict in the Middle East again returning soldiers with fewer limbs than when they left, the defense department began sponsoring more advanced research into prosthetics.

Prosthetic arms, hands, and legs today are essentially robots that interface with the body. These machines are strapped on and then slowly they learn what signals from the body or the even the mind of the user mean. Advanced prosthetic legs learn the gait of the user and are able to match it and use their own actuators to again put a spring in the step of the person. The most advanced prosthetic arm, the LukeArm built by DEKA, is able to read nerve signals from the user's brain and behave as a normal arm would. But this arm can only interpret a few commands, so the user does not have enough control to perhaps wiggle a single finger.

Currently most of these smart robotic prosthetics are still too expensive for widespread deployment. But since it is only the intelligence of the machines which is holding them back it is really a question of how long Moore's Law will have to operate before they can be affordable and more perfect.

From a business standpoint prosthetic devices for the handicapped is a very limited market. But the opportunity remains in simply enhancing normal people through robotic upgrades.

Such upgrades include exoskeletons, like the Iron Man suit which can protect workers from being strained, or assist the elderly in remaining independent. All such technologies are derived directly from the work done with amputees. Exoskeletons and simpler augmented systems are just robots you wear.

It is conceivable that as prosthetic devices improve, you may see surgical procedures which replace any biological part for a synthetic one. Though there might be questions about the viability of such technologies when they are often obsolete after a short period of time. But it is still something to consider. Robots may not change society by being separate from humans but by actually becoming a part of humans. Almost a digital mechatronic evolution of the human race.

What the Future Holds

On the technological side of things robots need to be smarter. That is really the primary challenge. The ability to make robotic bodies has existed for some time, though expensively. But the intelligence of robots is miserable because they are governed by a few basic equations called control algorithms which are really not very intelligent. The people who make these algorithms are very sharp but an algorithm rarely has the ability to, on a whim, learn to play the piano the way a human would.

Additionally, robots are a challenging technology because there are really no standards in place. In biology, cells, the basic building block of life, are built on DNA. There is a structure. And certain cells have certain functions. The neuron in a human is not that much different from a neuron in a frog.

But robots have no such foundation. Each robot is a Frankenstein monster compared to all other robots. It may have a brain that uses a neural network and his body is an octopus. Or its brain may be a laptop computer and its body is a Segway. The vast diversity of technologies and designs in robotics is a challenge because there is no rock to build on, only sand.

The industry really needs to work on ways of created robot DNA. Computer codes, languages, or parts, which nearly every robot or roboticist uses. Today the closest thing that exists to perform that function is a flowchart that says the motor is connected to the brain and the brains sees through a sensor. That is the current DNA of a robot, there is no definitiveness to it.

If I were recommending a project to young engineers I would suggest that they try to build such DNA. Perhaps a program that can universally identify what kind of a body a robot has. Or perhaps they can generate a better flowchart than what exists now.

Robotics is a great engineering discipline because it is so multifaceted and open-ended but it gets in its own way when it can't build on itself.

Many in the industry today compare robots to the personal computer. They are just on the edge of having the abilities needed to be useful to the average consumer. Personally, I disagree. I rather believe that robots are following a similar story to that of the telephone.

Just like the telephone robots have been created and are only understood and utilized in a technological or industrial setting. Entrepreneurs and average businessmen have not recognized their potential, just as what occurred with the original telephones built by Alexander Graham Bell. But slowly robots have gained wider acceptance and are now becoming something of a luxury item, in the form of robotic vacuum cleaners, to the slightly upper class who recognizes their use. The IRobot Roomba I would say is very much like a car phone or the old brick cellphones.

At this point the flip phone of robotics has not been created. The robot that everyone can afford, use, and learn to thumb-type on has not been created. This only has to be a robot that performs some task in everyone's lives that no one can do without. Again this could be compared to the Roomba but the Roomba is too narrow not everyone has to have a robot vacuum like everyone had to have a cellphone. So that step in the robotics industry has to be made.

But for robots to truly be phenomenally successful they must do what the smartphone did and become a multitool. Smartphones combined internet, cameras, a phone, and e-mail all into a single device.

Robots are still too dumb to perform multiple functions. Many in the industry believe that robots should be made function-specific. That is, you have a robot to vacuum and a robot to cook and a robot to walk the dog, etc. But that situation is like having a camera, a laptop, and a cellphone in your purse when all those devices should be wrapped into one.

The smartphone is truly the pinnacle of personal computing because it has combined tools more than any other product division. Or at least it has done so up to this point in time. Robots are still pre-flip phone in their existence. But that means that there are opportunities for entrepreneurs and inventors to get into before the industry explodes. And that explosion will likely occur in the next ten years. Then the smartphone of robots will arrive, the singularity will occur, and the human race will be replaced. At least that is what will happen if you are a dooms-dayer.

Personally I believe that robots will come to not replace jobs but to fill the millions of jobs left undone by people today because we simply do not recognize them. Whether intelligent machines will agree to perform such jobs ... that is a question that is not for this book. But robots will likely be like the apps on a smartphone. They will create an entirely new industry. Robots will be created with such variation that they will fill every pocket of our lives.

About the Author

Gabe Bentz grew up on a cattle ranch in eastern Oregon. He studied mechanical engineering at Embry-Riddle Aeronautical University. He now operates a robotics start-up in Boise, ID working to create a real robotic home butler.